What in the World
is a
Sasquatch?

A Review of Existing Evidence

by Phillip E. Buser

DEFIANCE PRESS
& PUBLISHING

What in the World is a Sasquatch?

First Edition: 2024

Printed in the United States of America

10 9 8 7 6 5 4 3 2 1

DEFIANCE PRESS
& PUBLISHING

ISBN-13: 978-1-963102-37-6 (Paperback)
ISBN-13: 978-1-963102-36-9 (eBook)

Cover photo credit: Ian Woodcock

Published by Defiance Press and Publishing, LLC

Bulk orders of this book may be obtained by contacting Defiance Press and Publishing, LLC.
www.defiancepress.com.

Public Relations Dept. – Defiance Press & Publishing, LLC
281-581-9300
pr@defiancepress.com

Defiance Press & Publishing, LLC
281-581-9300
info@defiancepress.com

Dedication

This book is dedicated to God in Heaven for giving me the wisdom to understand the evidence, the courage to put everything on paper, and the peace of doing His will.

I want to thank my wife and family for supporting me throughout this process; for letting me stay up late to write; and for dealing with my weird notions while researching various topics. They have endured, and for that, I am extremely grateful.

I also want to thank my friends, Pastor Edward Gardiner, Pastor Ilya Parkhotyuk, and Pastor Frank Harvey Logan who have continued to encourage and help me along on this journey.

Table of Contents

Chapter 1 | Introduction

There have been hundreds of reported sightings and encounters of Sasquatch across North America. Many reports come from ordinary citizens. Others come from police officers, civic officials, and people with a professional reputation on the line. But the one thing these people have in common besides Sasquatch is that they are willing to be ridiculed by their peers for saying out loud what they encountered. It's hard to imagine that all these people would make up a story in order to be shamed by others. These people encountered 'something,' so what *did* they encounter?

When I was a kid growing up in the 1980s, the term 'Bigfoot' had become mainstream and most everyone knew it referred to a large-footed creature hiding in the woods of the Pacific Northwest. Before the term 'Bigfoot' became part of culture's lexicon, reported encounters described a wild man, a hairy man, and even a forest monster. The descriptions often were similar to the famed

'Abominable Snowman of the Himalaya' that was popularized in Nepal in 1921. The original explorers that attempted to climb Mount Everest told reporters about seeing large footprints in the snow during their climbing expeditions. These initial reports to the Western World were then supported by Sherpa folklore of the Yeti. The stories of the Yeti began to inspire reports of Sasquatch encounters as the abominable snowman of the Americas.

The term Bigfoot was originally coined by a pair of reporters in northern California in the late 1950s. The original rise in popularity occurred in large part due to the Patterson-Gimlin film which became public in 1967. Remember the grainy, jarring video of Bigfoot walking across a rocky creek bed in Northern California? With this lone piece of video footage, a Bigfoot, or now more commonly referred to as Sasquatch, was still considered something of a myth—a little more real than unicorns, and about as likely to be seen as the Loch Ness Monster. I also grew up under the notion that the Sasquatch was seen by extremely few individuals in the most remote parts of the Pacific Northwest. And the odds of being able to see a Sasquatch was about equal to the odds of seeing the Loch Ness Monster.

When I was in college, I majored in Wildlife Biology. I would occasionally hear other students and professors talk about the myth of Sasquatch. From their skepticism, it became clear to me that the scientific community did not accept the possibility that Sasquatch was anything more than a fairy tale. Thus, from that influence, I accepted the same notion.

It wasn't until I was in graduate school that I began to hear stories of sightings occurring in other parts of the country and throughout North America. I heard about the 1970s documentary/movies *The Legend of Boggy Creek* and *The Legend of Bigfoot*. I also learned that sightings occurred at a frequency much greater than the lottery-jackpot winning chances I had been led to believe. At this point I began to take a more curious interest in the subject matter of Sasquatch. However, the amount of publicized information regarding Sasquatch was still rather limited. There would be the occasional 'Bigfoot Special' on various television shows, but that was still before everything imaginable could be found on the internet as is the case today.

I would have to say that it wasn't until the television show *Finding Bigfoot* began airing on television in 2011, and the sixth season of *Survivorman—Bigfoot*, that more information became readily available and easier to access for someone interested in learning about Sasquatch. Websites, podcasts, documentaries, docuseries, and television specials had newfound opportunities for people to share their experiences, information, and theories. The internet had grown in popularity and availability, so that people can 'surf the web' at any time. And streaming services began to provide an economical method of viewing a wider variety of documentaries and special interest shows. I can't fathom a guess as to how many hours I have spent watching, listening, and reading various media outlets about a Sasquatch subject. From all that, I found that the various theories which attempt to answer the glaring question, "What is a Sasquatch?" and the attempts to argue support for these theories, tend to come up short, or they blatantly leave massive gaps in their evidentiary analysis. A number of great books argue the existence of Sasquatch and detail the physical evidence to support that Sasquatch is in fact something real. Books like *North America's Great Ape: the Sasquatch* by Dr. John Bendernagel, and *Sasquatch - The Apes Among Us* by John Green give compelling evidence to conclude that Sasquatch isn't just a myth or fairy tale. And great websites and podcasts highlight many of the various encounters that people have had with Sasquatch. But what *is* a Sasquatch?

The search for information about Sasquatch has led to the rise in a new branch of science termed cryptozoology, which is the study of species not yet fully recognized. The scientific community has not fully recognized Sasquatch because there is a lack of evidence. There is not a deceased body to study in a lab. There is not a captured specimen in a zoo for researchers to observe. No one like Jane Goodall and Diane Fossey has been able to live among a group of Sasquatches to learn their behaviors. Therefore, Sasquatch is often referred to as a cryptid—a species that is not yet fully recognized by the scientific community.

Other creatures around the world have characteristics similar to the Sasquatch—a large, hairy human-like being that walks on two feet. Most notable is the Yeti of the Himalaya region. But there is also the Russian Almas, or Almasty. Reports of the Yeren occur throughout China. In Southeast Asia, the Orang Pendek is the hairy wildman that people encounter. In Iraq and the Middle East, people report encounters with the Dev (short for devil—frightening!). Even in Australia, reports of people encountering the Yowie which has its roots in Aboriginal history.

I have been a professional wildlife biologist for over twenty years. I have been an avid outdoorsman my entire life. And I am truly fascinated with the discovery process of knowledge that is science! I am not a Sasquatch hunter per se, as I have not gone out in the woods to specifically try to find Sasquatch. I have had some unexplainable experiences while out in nature, and I have found Sasquatch to be interesting subject matter on multiple levels. If you want to call me a skeptic, so be it. In the following chapters, I will give my critical review of the primary theories for the existence of Sasquatch, the support for each theory, and the glaring gaps in evidence for those theories.

This book is not intended to read like a doctoral dissertation, but rather a plain, practical discussion of the known evidence regarding Sasquatch. I want everyone that reads this book to be able to discern for themselves what a Sasquatch is, or isn't. Whether you agree with my analysis or not is irrelevant. The point is that you can follow the scientific method for yourself:

Make observations—What shows have you watched? What books have you read? Have you had experiences in the woods? Can you draw from any of these experiences?

- **Ask questions**—Does that make sense?

- **Formulate a hypothesis**—Do you agree with other theories, or do you have your own theory?

- **Test your hypothesis / gather evidence and data**— Continue researching.

- **Draw conclusions based on your evidence**—Are your results repeatable for others?

- **Share your conclusions with others**—Let others critique your process and findings.

A scientist's greatest role is to critically review the published results of others.

Then, and only then, can you critically look at any point of view regarding Sasquatch (and any other topic); know the pros and cons of the various theories; and make an informed decision for yourself.

Chapter 2 | The-Sasquatch-Isn't-Real Theory

Before I delve into the theories about what Sasquatch is, or isn't, I want to address the theory that Sasquatch isn't even a real thing.

The first aspect of this theory is that many Sasquatch encounters are mis-identification. Individuals that support this theory often claim that persons who have an encounter, simply encountered something else—primarily a bear, or possibly another person. Now I'm not going to debate how many reported encounters are truly encounters with Sasquatch, and how many are possibly mis-identification. But regardless of what that percentage may be, a certain number of encounters have other situational information. For example, consider those encounters where the Sasquatch left footprints on the ground, or where the Sasquatch broke a tree branch or something else at a height and in a manner that would be impossible for a bear or a human being. Needless to say, many reports of Sasquatch encounters have corroborated evidence to support the encounter was something other than a bear—or more notably something *like* a Sasquatch.

Another aspect of the misidentification theory that critics point to is that of cultural influence. The term Bigfoot has been in the cultural vocabulary since the 1950s, and the image of a typical Sasquatch has been broadcasted and reproduced since the 1960s. Every year, the subject of Sasquatch gains popularity because of the cultural influence that 'Sasquatch' has on everyone in our society; that influence can subconsciously influence people. Therefore, if someone is driving at night, walking in the woods, camping, or whatever, and they see something in the distance, or out of the corner of their eye, that cultural influence may cause the individual to 'jump to the conclusion' that what they saw was a Sasquatch. Whether or not they did see a Sasquatch, they convince themselves that what they encountered 'must have been a Sasquatch.'

Yet another aspect of this theory is hoaxing. This is the notion that people intentionally fabricate encounters and evidence. Quite a number of Sasquatch encounters have been dismissed by the investigators—whoever they happens to be. A number of Sasquatch footprints have been proven to be false. I previously mentioned a few books that detail the differences between footprints that are a hoax and footprints that are likely from a real Sasquatch. A few individuals have publicly claimed to be part of a Sasquatch hoax. They made their claims as a way to get public recognition and notoriety. However, the growing number of reports from across the country are too numerous and too varied for the claimed hoax originators to have caused all of them. So, for these other reports of encounters with Sasquatch, the question remains: What did they encounter?

The last aspect of this theory is the lack of substantial physical evidence. The question that always comes up is: Why hasn't anyone found a body? Along with this question is the issue that no one has found even part of a Sasquatch body, or bones, or fossils. These relevant questions need answers, and I will discuss these issues in the following chapters.

Chapter 3 | The-Undiscovered-Species-of-Ape Theory

One theory to explain the existence of Sasquatch is that Sasquatch is an undiscovered species of primate. Perhaps it is another species of ape, or the evolutionary descendants of Gigantopithecus—an extinct species of gorilla that was nearly ten feet tall.

I have often heard Occam's Razor as the primary supporting argument for this theory of Sasquatch. Occam's Razor states that the simplest, most logical explanation, with the least amount of assumptions, is usually the correct explanation. I will agree that this theory is the simplest explanation because another species of animal is also the easiest for most people to accept. However, it is far from the most logical, and here's why.

One can find various stories, local newspaper clippings, etc. of Sasquatch-like beings in North America dating back to the Pilgrims, and further back when you add Native American stories. However, Sasquatch didn't become a household character until the before-mentioned Patterson-Gimlin film became public in 1967. I should note that this video has always been controversial because some people doubt its authenticity. Regardless, it has now been fifty-five plus years since the release of this video. Despite the unknown hours that numerous people have spent searching for this hide-and-seek champion, the most that science has to show for the effort is a growing collection of cast footprints and a few strands of 'unknown DNA' hair. That is the tangible evidence, the evidence that people can touch, look at over and over again, and critically analyze for every shred of information that can possibly be gleaned from it. This doesn't take into account the circumstantial evidence of odd and unusual (and sometimes frightening) sounds that people hear in the woods, or even the blurry, dark blobs caught on video moving through the woods, which is pretty much every other video that anyone has offered as potential evidence for Sasquatch. I point this out as a contrast to the discovery of the Mountain Gorilla of Africa. African wildlife has been known throughout the world for centuries because of trade among the various African tribes, the Egyptian kingdom, the Roman Empire, and everyone in between. However, it wasn't until the British Empire was in full colonization mode of Africa in the 1800s that rumors of a large primate hiding in the jungles of Africa finally made their way to Europe and the 'civilized' world.

Surprisingly enough, it took less than twenty years to have a living, breathing specimen in the London Zoo. Granted, the acceptance of the gorilla as a separate species took several more years of study. But, nonetheless, a gorilla was captured so that additional study could be done. And that was using nineteenth century technology. That technology included ropes and pulleys, bamboo cages, and hand-dug pits for traps. There were no thermal imaging devices, trail cameras, motion sensors, etc. to help locate and trap a gorilla. Doesn't it seem odd that after fifty-five plus years of searching for Sasquatch, we have only one old video, some footprint casts, and a couple strands of hair? The British were able to capture a live specimen of the gorilla in less than twenty years more than a century earlier. How odd.

I would like to mention that some recent Sasquatch researchers have presented rather interesting videos regarding Sasquatch. The videos include some shots that zoom in on the faces of several Sasquatches hiding in the brush. I have not met any of these researchers and know little more about them and their videos. I would also like to mention that there is a swirl of controversy regarding these videos. I, and many others, find it odd that individuals were able to get into a position that allowed capturing the faces of Sasquatch on video, but were unable to capture the full body of those Sasquatches either entering or leaving those hiding places.

As I mentioned in Chapter 1, as a kid, I was under the notion that Sasquatch was only hiding out in the deep forests of the Pacific Northwest. And it wasn't until I was in college that I began hearing stories of Sasquatch encounters in other parts of the country. So, if populations of Sasquatch are located throughout North America, and not isolated to the deep forests of the Pacific Northwest, wouldn't that provide more opportunities for people to find more significant evidence, carcasses, etc.? But yet, here we are minus fifty-five years and counting with only a grainy video, cast footprints, and a few strands of hair to show for our knowledge of the hide and seek champion.

The next area of preponderance is the lack of skeletons and a fossil record. One of the most prominent outlets for those interested in the Sasquatch phenomenon is the Bigfoot Field Research Organization (BFRO). On the BRFO website, they report Sasquatch sightings in forty-nine states (no reports from Hawaii) and nine Canadian Provinces. The website addresses the lack of skeletons, carcasses, etc. by claiming that scavengers and natural processes of decay consume and recycle any deceased Sasquatch quickly, which is why carcasses/body parts are not found.

I have heard several biologists make this claim. Additionally, in regions where the climate is wet enough to grow trees, the decay process speeds up, making it even more difficult to find a carcass or skeleton. That is true. However, sightings are reported in areas with dryer climates. So why haven't the remains of those Sasquatches been found? If you're an avid outdoor enthusiast, I'm sure you have stumbled across a bone of some dead animal at some point in time. Not all animal parts decay before someone can find them.

There is a large segment of the general hunting community that goes out in the woods during the winter/early spring season to look for antlers shed by deer, elk, and moose. These animals annually lose their antlers, then grow a new set of antlers during the next year. These people are literally scouring the ground looking for animal parts. Why haven't we heard a report that even one of these individuals has found a deceased Sasquatch? If you can find a bone of a deceased deer, elk, etc., then there is no reason why a Sasquatch bone has not been found.

Why do people find dinosaur fossils all over the world? Dinosaurs lived even longer ago than Sasquatch. People find fossils and remains of other extinct animals in North America and beyond. The reason is that not every carcass gets 'recycled' the same way. Some get preserved for a later time. Some fossilize. But apparently that has never happened to a single Sasquatch ever in the history of North America.

The other flaw in this argument is that it seems to assume that Sasquatch has inhabited North America for only a short period because the Sasquatches that have lived (and died) in North America haven't yet had time to become fully fossilized or preserved. Proponents claim that all Sasquatches that have died in North America have fully decayed. We know that many species of wildlife once ran wild across the continent, but no longer exist: mammoths, saber-tooth tigers, the North American cheetah, and many others. We know this because we have found preserved remains of these species. We assume that Sasquatches have been on the continent for a long time since Native American cultures included them in historical accounts. That adds to the amount of time during which a number of deceased Sasquatch carcasses could have been preserved.

Maybe Sasquatch carcasses decay at a faster rate than other animals. Perhaps the odor of a Sasquatch is more inviting to vultures, worms, bacteria, etc. After all, a common 'description' for

Sasquatch mentions the presence of a strong, pungent odor. But biologically speaking, that is highly unlikely.

It's not like we haven't had opportunities to find their remains. A Sasquatch may be elusive and good at hiding when they are alive. However, once the Sasquatch dies, it can't get up and go hide again if some hunter, hiker, developer, etc. stumbles on its final resting place. The next time you have a free moment, pull out an atlas of the United States. As you flip from page to page, state to state, pay attention to the roads, highways, cities, towns, streets, and all the other infrastructure developments that are identified on the maps. Bear in mind that it doesn't detail the houses, stores, offices and other buildings of every community in this country. For all of these structures to be built, a foundation must first be established. And to set the foundation, someone has to do a lot of digging—physical labor type digging. People have argued that if someone uncovered a Sasquatch bone or skeleton during excavation work, that person would cover it back up; act like they never saw it, and proceed in a normal way to avoid potential delays in the construction project. Yes, that is possible. But my experience is that people have a strong sense of curiosity. More often than not, they want to learn the origins of what they uncovered. That initial curiosity tends to outweigh a possible delay of a construction project, especially if the finders are about to be credited with a major scientific discovery. People like to know what's going on around them. That's human nature. Yet, no one has brought forth a carcass or skeleton of anything that could be argued to be a Sasquatch. It would be different if Sasquatch skeletons were somewhat commonly found. Then, finding 'another' skeleton might become burdensome to the progress of a construction project. But we're not at that point yet.

I mentioned the documentary *The Legend of Bigfoot* in Chapter 1. This 1975 documentary helped increase the popularity of Sasquatch across the country. I mention this movie again because one claim in the documentary indicated that Sasquatch carry the remains of the deceased to certain burial grounds far north, near the Arctic Circle. Now I can't rule that out for any Sasquatch that resides in northern Canada or Alaska, but what about all of the Sasquatch living in the continental United States? After all, have you heard of any Sasquatch encounters that involved a Sasquatch carrying another dead Sasquatch on a northern migration? I have not. The documentary also suggested that Sasquatch are migratory. The documentary suggested that these are long migrations from the southern United States to Canada. But how does that explain reported sightings occurring in all parts of the country and Canada, during every season of the year? For Sasquatch to make short migrations, such as from high mountains to lower valleys is much more plausible. Even from the southern part of a state to the northern part of a state is still within the realm of possibility. But to make long, continental migrations seems much too unlikely.

Let's get back to that atlas. When you looked at the atlas showing the highways, by-ways, and roads across this country, you may have wondered about roadkill. That is a Sasquatch getting hit by a moving vehicle. Given the large size of a Sasquatch, a vehicle collision may not always result in the death of the creature and provide a carcass for people to see. But there should be a few casualties. The argument that I hear more often to discredit the possibility of a roadkill Sasquatch is that *Sasquatch is elusive.* Thus their secretive nature makes them avoid busy roads. But as I mentioned earlier, if you inquire into even a few encounters reported across the country, you'll notice that a large percentage of the reports are from people who are driving. Whether it's a dirt road through the countryside, or a highway late at night, you'll realize that even a secretive Sasquatch can't avoid being seen on the thousands upon thousands of miles of roads in this country. Shouldn't that account for some number of roadkill? After all, even the elusive mountain lion and other secretive species of wildlife occasionally get hit on a roadway.

By the way, if Sasquatch is an elusive creature that lives a life of secrecy, how is it that the BFRO alone has collected reports of sightings in forty-nine states and nine provinces?

There are also a number of smaller organizations, a handful of museums, and local newspapers in several states that collect reports. Isn't that unusual for an 'elusive' creature? Even though this BFRO has been around since 1995, their members have investigated sightings and reports from many years previous to their inception. To their credit, The BFRO acknowledges only reported sightings and encounters that their members have been able to verify as credible. The BFRO claims some level of credibility for the report. Looking through the reported sightings and encounters just in the United States, I noticed something peculiar. The reports are listed by state and by county. It just so happens that there are more than the 'freaky coincidence' of sightings and encounters near major metropolitan areas—and not just the somewhat expected Seattle and Portland, as they are located in the Pacific Northwest, the historic hotbed of Sasquatch sightings.

State	# of Listings	Most Recent Report	Last Posted	State	# of Listings	Most Recent Report	Last Posted
Alaska	22	12-2022	5-2022	Montana	53	7-2020	4-2020
Alabama	102	2-2023	2-2023	North Carolina	106	7-2022	7-2022
Arkansas	111	3-2023	9-2022	North Dakota	6	12-2010	8-2005
Arizona	85	6-2021	6-2021	Nebraska	15	1-2019	5-2018
California	461	12-2022	10-2022	New Hampshire	17	5-2021	7-2016
Colorado	130	7-2021	6-2021	New Jersey	75	4-2022	6-2019
Connecticut	22	12-2022	7-2022	New Mexico	43	7-2021	7-2020
Delaware	5	9-2013	11-2012	Nevada	9	4-2009	2-2005
Florida	339	3-2023	3-2023	New York	120	11-2022	10-2022
Georgia	140	12-2022	12-2022	Ohio	320	3-2023	2-2023
Iowa	76	4-2018	6-2018	Oklahoma	111	10-2022	8-2022
Idaho	103	9-2022	11-2021	Oregon	257	12-2020	8-2020
Illinois	302	3-2022	11-2021	Pennsylvania	126	2-2023	12-2022
Indiana	82	9-2020	6-2020	Rhode Island	5	12-2011	11-2011
Kansas	49	5-2021	5-2021	South Carolina	57	8-2022	8-2022
Kentucky	115	12-2021	11-2021	South Dakota	19	6-2020	6-2019
Louisiana	44	12-2020	7-2018	Tennessee	104	10-2021	6-2021
Massachusetts	37	9-2020	8-2020	Texas	254	3-2023	3-2022
Maryland	35	11-2022	11-2022	Utah	73	9-2022	9-2022
Maine	21	1-2023	8-2022	Virginia	86	2-2023	11-2022
Michigan	225	2-2022	2-2022	Vermont	10	11-2019	8-2019
Minnesota	77	12-2022	10-2022	Washington	710	2-2023	8-2022
Missouri	166	7-2022	7-2022	Wisconsin	106	2-2023	9-2022
Mississippi	24	1-2022	1-2022	West Virginia	106	10-2022	7-2017
				Wyoming	28	5-2010	3-2010

Reports of Sightings

There are considerable reports around Cleveland, Detroit, Fort Worth, Kansas City, Los Angeles, Louisville, Phoenix, Pittsburg, and Salt Lake City. I'm sure the argument would be made that the occasional encounter is inevitable with the expansion of these cities into previously undeveloped Sasquatch habitat and so many people moving in. There is the statement that, "Eventually someone is going to win the lottery." So how does a secretive, elusive creature hide out in the vicinity of huge cities; let itself be seen by at least a few people in those cities, but *not* leave evidence beyond the occasional foot print?

Now, let's look a bit closer. I'm not a professional statistician, but I do know a thing or two about statistical analysis. After all, I took three statistics classes in college (one in graduate school). As I previously mentioned, the BFRO website provides Sasquatch encounters for forty-nine states, broken down by county. Coincidentally, the United States Census Bureau provides human population data for each state (also broken down by county). And that data is available on their website. So I took a little time to copy and paste the Census Bureau data into a spreadsheet along with the observation data from the BFRO website. Once I combined the county-by-county data from the Census Bureau and the BFRO websites, I ran a linear regression analysis. This allowed me to see if a significant relationship exists between the number of humans in a county and the number of Sasquatch encounters that occur in the same county.

I understand that some encounters involve historical reports, and that the current population data of a county has changed since some of the observations. This analysis did not show any direct significant relationship between the number of people in an area, and the number of Sasquatch encounters in the same area. But that is not where this story ends. I sorted the data by observation total. This is when I noticed something interesting. According to the US Census Bureau, there are 3,108 counties in these 49 states, not counting Hawaii. The average human population for those counties is 98,109 people. And according to the BRFO, there have been Sasquatch encounters in 1,571 of those counties. Now get this: It just so happens that ten of the top twenty counties for Sasquatch observations have a human population over 100,000. And going a little further, twenty-nine of the top fifty counties, and fifty of the top one hundred counties for Sasquatch encounters have a human population over 100,000. These counties are not just in the Sasquatch hotspots of California, Oregon and Washington. They include counties from Illinois, Ohio, Oklahoma, Florida, Utah, Minnesota, Texas, Maryland, Massachusetts, and Connecticut! And going yet a bit further, ninety-two of the top 200 counties, and 179 of the top 500 counties for Sasquatch observations have a human population over 100,000 people! I present this analysis for this reason: Given the abundance of Sasquatch encounters within heavily populated counties, why is the amount of physical evidence still so limited? Logic suggests that there should be much more physical evidence of Sasquatch in areas inhabited by both Sasquatch and humans. Again, I realize that probably many of those counties did not have that size of human population when some of those encounters occurred. And someone might argue that as people continued to move into the county, the Sasquatch that were living there simply moved on to less dense areas. But as roads and houses and stores, etc. were built for the expanding human population, why haven't the remains of previous Sasquatch that lived and died in the county been found?

I want to get back to the road observations. As I previously mentioned, a large percentage of reported Sasquatch encounters involve someone seeing a Sasquatch crossing the road in front of them. And a large percentage of those reports occurred at night. And again, I have often heard people claim that Sasquatch are extremely elusive, which is why there isn't more evidence of them. When I was in graduate school, I conducted a research project on roost-site selection of wild turkeys. Night after night, another research assistant and I would trek around in the woods searching for wild turkeys sitting on their roosts. We used radio telemetry and night vision goggles to make it less like looking for a needle in a haystack in the dark. Along with that, I was a certified law enforcement officer (game warden) for ten years. I mention all of this because I have spent a lot of time hiking around in the woods at night. And in doing so, I know how easy it is to see and hear a vehicle driving down a country road in the middle of the night. A vehicle has bright headlights and a noisy engine that makes it easy to see and hear. So riddle me this: If a Sasquatch is an elusive creature that doesn't want to be seen, why do so many road crossing observations occur at night? I mean really, if a Sasquatch truly wanted to remain hidden, it could easily wait until the approaching vehicle drives by, and then cross the road. No one would know that a Sasquatch was nearby. But no, it crosses the road in front of the vehicle making itself visible to the driver and other occupants. That doesn't sound like the behavior of a creature that wants to be hidden. What is more interesting is that from these sightings, there still are no reports of roadkill Sasquatch. Other animals are observed crossing roads at night, and some number of those animals get killed, but not Sasquatch. They seem to always know exactly when to cross the road in order to be seen, but not be hit by a vehicle. That's odd.

I hear many 'experts' claim that Sasquatch is an apex predator. This means that Sasquatch is at the top of the food chain. Some animals, although they are predators themselves, are also potential prey for other larger predators. An apex predator is not a prey species for any other creature. Grizzly bears, mountain lions, and wolves are other apex predators. Humans are also considered apex predators. Sasquatch are large creatures and reported to be fast and strong. So it is easy to see why Sasquatch are considered apex predators. No one has been able to do a complete analysis of a Sasquatch diet; however, a number of witness reports have stated that they saw a Sasquatch kill a deer and eat an animal carcass. So there is some evidence that Sasquatch are carnivores (meat eaters), if not omnivores (eat plants and animals).

Here is my dilemma: As I have said, there are reported sightings of Sasquatch across North America. In the United States, east of the Rocky Mountains, you are unlikely to find grizzly bears, mountain lions, or wolves. At least not in substantial numbers. Now, when populations of prey species, like deer, don't have natural apex predators (outside of humans), that population will grow out of balance with nature. That population will have negative impacts on their habitat. Individuals will venture into areas where they wouldn't go otherwise. And they will exhibit an almost carefree attitude because they are not afraid of anything bad happening. When there are apex predators around, things change. Of course there is the obvious result, that the population will decrease because of being preyed upon, but the other significant change is that the prey species become more elusive. Individuals are more cautious about where they go and what they do.

A perfect example of this is the response of elk in Yellowstone National Park after the reintroduction of wolves. Prior to the reintroduction, the elk population was above the park's natural carrying capacity. The natural habitat was not able to sustain the number of elk. After the reintroduction of wolves, the population of elk decreased. They were not wiped out by the wolves, as many people thought might happen. But the elk did change their behavior and their patterns. Today there are healthy populations of both wolves and elk, and other wildlife species as well.

Now, let's apply this to Sasquatch. Again, there are reported sightings of Sasquatch across North America. What are the populations of deer like in states east of the Rocky Mountains? In most regions, they are high. If you live in one of these states, go for an evening drive through the countryside. Slow down and take a good look at any clearing or field you drive by. You are likely to see a large herd of deer standing around, grazing to their heart's content. And at the next clearing or field, you will see the same thing—more deer.

Hunting regulations are another easy indicator of the current deer population in a region. You can look online at the hunting regulations for your state. If you scroll through the regulations, look to see how many does (female deer) an individual hunter is allowed to harvest in a single hunting season. It will be more than one, and the actual number will probably be surprising. So, if Sasquatch are populous enough to frequently get seen all across North America, and not just in extremely remote areas, and they are a predatory species of primate, apex predator or not, then we should not be seeing the overabundance of prey species (particularly deer) in the same vicinity as Sasquatch. Especially given their large size; the caloric intake has to be enormous. Yet that is exactly the opposite of what we have going on across much of the country. Deer numbers are high in most of the country, and their behavior is typically carefree. The predator-prey population dynamics are inconsistent with Sasquatch being just another primate.

The most prominent feature of a Sasquatch is their overly large feet—hence the name Bigfoot. The large feet are biologically consistent with what a large bi-pedal creature should have. And several good books have been published regarding the biological adaptations of Sasquatch feet. There is Dr. Grover Kanz' book *Bigfoot Sasquatch Evidence* and Dr. Jeff Meldrum's *Sasquatch: Legend Meets Science*, to name a couple. They have analyzed countless numbers of the growing collection of footprint casts, and identified several key features that are evolutionarily essential for a creature as large as Sasquatch. And I don't have anything against their analyses. I agree with them wholeheartedly. This is what I don't understand:

A Sasquatch has feet that are evolved to carry the weight of a 500-800-lb. creature, as they *walk* on two feet. And I want to emphasize the walking part. Their feet are adapted and specialized for walking, not running. Now, I'm not saying they are incapable of running as humans can run (and some are rather fast) when we *need* to run. But creatures evolved for running fast do not have large flat feet. They are running on their digits (fingers and toes). Horses and other ungulates even run on their enlarged fingernails! That is the accepted biological theory of evolutionary adaptation for swift-footed running animals.

Now to directly contrast what is known about running animals: In every Sasquatch encounter I have seen or heard involving the Sasquatch running, the eyewitness has emphasized how fast the Sasquatch was. The eyewitnesses say things like, "It couldn't have been human because humans don't move that fast," or "It was so fast, it's like it was floating," and, "I've never seen anything move that fast." I find this rather remarkable for a creature that has definitely 'evolved' for walking.

As I have mentioned earlier, the descriptions of Sasquatch are 6' to 8' tall (or taller), and 500-800 lbs. (or larger). I occasionally hear about in Sasquatch encounters where the Sasquatch was climbing around in a tree. At face value that doesn't seem out of the ordinary. People suspect that Sasquatch is a primate, and other primates are quite adept at climbing trees, and even humans can climb trees. Growing up, my friends and I often climbed trees, and occasionally fell out of trees. However, the largest gorillas (mature silverbacks) rarely climb trees because trees are unable to hold their weight. Now with Sasquatch, they are two times larger (or more) than the largest gorillas. When I was climbing trees as a kid, I learned quickly that a tree branch had to be pretty sizable to hold my weight. I can't imagine seeing a Sasquatch that is twice as big as a gorilla, and four or five times larger than I was when climbing around in a tree. Those must have been select, strong trees to support the weight of a Sasquatch!

For many years I have heard people tell of their encounters with Sasquatch on podcasts, television shows, documentaries, etc. . . . Many people share a common thread—an overwhelming sense of fear during the encounter. In many instances, the sense of fear is so overwhelming that people describe themselves as being paralyzed. Where does this overwhelming sense of fear come from? I bring this up because it is far from a normal response. Most of the individuals who have made this claim are avid outdoor people. They feel comfortable being in the woods surrounded by nature. And this sense of fear is in stark contrast to encounters that other similar people have with other large animals in the woods. Let me explain. When an experienced hiker sees a bear on a forest trail, the hiker typically doesn't have an overwhelming sense of fear that leaves their body paralyzed. Even people who are not seasoned veterans of the hiking trails, and don't know the difference between a pine tree and a spruce tree, don't respond like that to animal encounters. I'm not saying that their heart rate doesn't go up. Maybe some extra adrenaline gets pumped into their blood, and they become extra aware of their surroundings. That would be a normal, biological response to encountering a large predator in the woods. An overwhelming sense of fear that leaves one paralyzed is not a normal, biological response, even though that is exactly what happens—and frequently. If Sasquatch is merely an undiscovered primate, why do people have an overwhelming sense of fear during their encounters? That is not normal.

This fear phenomenon doesn't occur only with humans. When the witness has been accompanied by a pet dog, nearly every encounter I have come across involves the dog whimpering, cowering, or even running away and hiding. This is quite contradictory to a dog's normal reaction to other animals. When a dog senses (hears, smells, etc.) another animal, it starts to bark. Even if the dog is small, it barks. Even if the animal is a bear, or a mountain lion, or some other large predator that could easily kill that particular dog, they still bark. But if the dog and its owner encounter a Sasquatch, they don't bark. They stay quiet and hide. That too is not normal.

Here is something else that isn't normal. Along with people having overwhelming episodes of fear and pets cowering or hiding, it seems that the other creatures in the woods take notice also. I have seen and heard about many encounters where the eyewitness stated that everything in the woods went quiet. People say that the birds stopped chirping, insects stopped buzzing, and the whole woods became so quiet that they could hear a pin drop. The eyewitnesses typically mention that they noticed the quietness right before their encounter with Sasquatch. After the Sasquatch walked away (and if the witness didn't flee), then the normal sounds of the woods returned. I have spent a lot of time in the woods throughout my life. I hunted and fished, hiked and camped, and my career involved working in the woods. Only a few times have I noticed quietness fall over the woods. A Sasquatch never showed itself to me during those times, and I cannot definitively say what caused the noises to stop including bird's chirping, but what I can say: That type of occurrence is definitely not normal!

As I have mentioned, one of the few pieces of evidence for Sasquatch appearances are plaster casts of footprints occasionally found throughout North America. Footprint casts provide evidence that a heavy body with large feet recently walked there, and did so in a relatively recent timeframe. It also provides clues to the overall morphology of Sasquatch. What they don't provide is evidence as to where Sasquatch is going or where they came from. I'll explain what I mean.

People who have found large footprints in the middle of nowhere or even in a nearby park, suspect Sasquatch made them. Most of those people have tried to follow those tracks and investigate where they lead. And they usually follow the tracks back to wherever they originated. Every eyewitness account that I know of regarding Sasquatch tracks, people have stated that the tracks started and stopped suddenly. What does that mean? People say each set of Sasquatch tracks seems to start in a random location; the tracks continue for some unspecified distance; then the tracks stop as randomly as they started. These eyewitness accounts also make a point of stating that whatever the footprint medium is (i.e., mud, snow, sand, etc.) that particular medium continued in each direction. That means that the tracks could have, and should have, continued in that direction. But they didn't. They just stopped (or started) for no apparent reason. It's not like the tracks started at the transition area of a rocky area to a muddy area. It was muddy or sandy or snow-covered etc. This rules out the argument that, "the tracks were only found in the mud because that was the only place that was muddy." The descriptions from the eyewitnesses convey a message that the Sasquatches that made the footprints were air-lifted in the starting point and then air-lifted out from the endpoint. How does a large primate plop down in the middle of nowhere, walk for some distance, then seemingly fly off? How odd.

The best piece of evidence for the existence of Sasquatch is the Patterson-Gimlin video. It was recorded by the eyewitnesses Roger Patterson and Bob Gimlin over fifty-five years ago. Cameras and video recording equipment have come a long way in fifty-five years. There are trail cameras with infrared lights, Bluetooth connectivity to smartphones, satellite connections, and so much more that people are able to set out in the woods. And they do! People are always sharing pictures and videos of people and creatures on social media and the internet. Either from their own property, or somewhere a little more remote—everywhere is seemingly a click away.

Remember the boom in homemade hunting videos? There was the rise in popularity of commercially made hunting videos, made by film crews and production teams. Why didn't they record any Sasquatch encounters or glimpses of Sasquatch doing Sasquatch things? Many entrepreneurial hunters began taking their own video cameras on hunting trips and recording themselves, or maybe filming their hunting buddy, as they stalked their intended prey in various locations across the country and around the globe including Sasquatch habitats. Many of these hunting videos show the hunted game animals moving from hundreds of yards away to within shooting distance, sometimes merely a few yards away. In these hunting videos, at whatever distance the animals are first seen, the clarity is good, and one can easily identify that animal. So if someone did encounter a Sasquatch at that same distance, it would be a video with enough quality to easily identify it as something other than the common wildlife species of that area.

GoPro cameras became all the rage and people were videoing their adventures of mountain biking, hiking, and every crazy stunt they could think of. How many hours were spent in remote areas, aka Sasquatch habitat, without a Sasquatch encounter being recorded?

Despite all the cameras and all the technology that has been deployed for the purpose of catching an unsuspecting creature on film, all of the hunting videos, all of the outdoor adventure videos that have been recorded in the past fifty plus years, nothing has been brought forth with a better image for a longer period of time than the Patterson-Gimlin film, which was approximately thirty seconds long. When someone does come out to say that they caught Sasquatch on camera, the image is so blurry or distorted that the Sasquatch appears to be a dark and blurry blob, at best. Why is it that with so many advances in camera and video technology over the last half-century, all of the man-hours spent

ing Sasquatch) that no one has been able to bring forth a better image or video of a Sasquatch? How odd.

So with the many advancements in technology, there should be an appropriate acceleration in the amount of and quality of evidence for Sasquatch. But that is not what the Sasquatch researchers are experiencing. The best evidence for Sasquatch is some of the oldest evidence, the Patterson-Gimlin film and some of the best footprint casts. This is the opposite of what is happening in every other field of research!

One more problem with this theory. It is the existing evidence. Don't get me wrong; there is a lot of evidence for the existence of Sasquatch—thousands of encounters and footprints left in the ground, the Patterson-Gimlin film, and some hair. What I am getting at is the inconsistencies. Here is what I mean. If a person wants to hunt ducks, they gather duck decoys, and a duck call, and then go out to a wetland. They set out the decoys; make quacking sounds with their duck call, and there is a good chance some ducks will fly by and give the hunter an opportunity to shoot them. One other example is the snow leopard. The snow leopard is one of the most elusive species of land animal. However, if you have the time and money, you can travel to the mountains of central Asia, climb up to the elevation that snow leopards prefer, and if you stay long enough, you will have a pretty good chance of seeing a snow leopard. And you could possibly even get some good photographs of one. In both of these examples, the hunter and the photographer will probably meet their goals because previous researchers have identified habitats and behaviors of ducks and snow leopards and that experience can be repeated by others with similar probabilities of occurrence.

Now let's apply that to Sasquatch. As one investigates the many Sasquatch encounters, many similarities become evident. The difference is the predictability. Let me explain. When wildlife researchers conduct studies on various species of wildlife, they collect data on a host of parameters. For example, when I was in graduate school researching roost site selection of wild turkeys, I collected data on the height of the tree; the height at which the turkey was roosting; the distance to the place on the limb where they were sitting; the density of the tree canopy; and the amount of ground cover around the tree. Wildlife researchers collect their data and then analyze it for correlations and probabilities of occurrence. Then their research is corroborated with other researchers. From all of that effort, others can go out and have a statistically significant chance of reproducing the same results.

With Sasquatch encounters, there are massive inconsistencies. First, when an encounter occurs, other people will often go to the same location, the same time of day, and see nothing. It is often claimed that Sasquatch communicate by striking trees with sticks; this is called 'wood knocking.' However, if you watch many of the videos on the internet about people going out and looking for Sasquatch, they try wood knocking to elicit some kind of response from Sasquatch in the area, and get nothing. Or even if you talk to other Sasquatch researchers, more often than not, they do not get any kind of response from their wood-knocking efforts. The times that a wood knock gets answered by a suspected Sasquatch are few and far between. And even a Sasquatch encounter is highly unpredictable. I've already talked about the lack of photographs. Getting photographs is harder than seeing a Sasquatch, or hearing a wood knock. This is not normal for any other species of wildlife. The behaviors of animals, whatever they may be, provide some level of predictability of occurrence. But not Sasquatch. That's odd.

You have probably heard someone say that the definition of insanity is doing the same thing over and over, and expecting different results. We as a society have been searching for Sasquatch diligently for over half a century. And by all accounts, we are no closer to proving Sasquatch is an undiscovered primate. So, if you believe that Sasquatch is a great ape wandering the wilderness, please re-read the first sentence of this paragraph.

Chapter 4 | The-Missing-Link-in-Human-Evolution Theory

The next theory I am addressing is that Sasquatch is not an undiscovered primate per se, but is actually the missing link in human evolution. It never went extinct. This theory hypothesizes that Sasquatch is an early species of human, such as a Neanderthal, that never went extinct. This theory claims that the Sasquatch human species went into hiding in the woods, as opposed to becoming an extinct early version of modern humans. By retreating and hiding, the Sasquatch groups isolated themselves from those that would eventually evolve into modern humans.

One major positive aspect to this theory is skeletons. Skeletons of old Neanderthals have been found around the world. It should be noted that sightings of creatures similar to Sasquatch have also been reported around the world. Unfortunately, no one is claiming to find skeletons of recent or modern era Neanderthals. All of the known skeletons of Neanderthals are reported to be over 100,000 years old.

It is easy to see the popularity of this theory, given the common body size and shape described in most Sasquatch encounters. It also would better explain the somewhat wide variety of body size and shape descriptions that people elaborate on from their encounters. The common body size and shape of Sasquatch encounters is that of a tall, heavy-set humanoid. But when you look at Sasquatch reports more closely, you find a wider range of descriptions. The variety of body size and shape descriptions that I am referring to is that the height ranges individually from six to twelve feet. The color of the body hair varies from reddish tan to brown to black. The face varies from being similar to a gorilla with large flaring nostrils, to having the face of a modern man. Sometimes the nose is flat, and sometimes the nose is pointed. For all species of animals, there are physical differences among individuals (phenotypes). However, the degree to which individuals are visually and physically different in the descriptions of Sasquatch is only comparable to that of modern humans. Few species of animals have the amount of diversity from individual to individual that is seen in humans and Sasquatch.

The Sierra Sounds is another bit of evidence that people use to support this theory. In 1971, Ron Morehead started recording sounds that he and others were hearing in the woods at a hunting camp deep in the Sierra Mountains. The following year Allen Barry joined Ron in the recording efforts. In 2008 they had the recordings analyzed by a retired Navy crypto-linguist. This individual had military training to identify hidden languages, and the specialist claimed that whatever was chattering away on the Sierra

Sounds recordings had the key characteristics of a discernable language. The ability to create a word-based language has been highly regarded as a skill only held by human beings, but apparently Sasquatch has that skill as well.

If this is the right theory for Sasquatch's identity, then I again have to ask: Why don't we have more evidence? Why don't we have more skeletons? I'm not implying that there are known skeletons of Sasquatch, but we do have at least some skeletons of Neanderthal. Why haven't scientists been able to locate Sasquatch skeletons and observe their behaviors? Why do people have an overwhelming sense of fear during an encounter? Why do their footprint tracks seem to start and end in thin air?

According to anthropologists, the evolutionary separation of apes and the earliest hominids occurred sometime around five million years ago. This is also when hominids began walking upright on two feet. From that time forward to approximately 200,000 years ago, the brain size of humans increased to that of modern humans. As the size of the brain was growing, the early humans developed stone tools, and began living in dwellings and campsites. Then, after developing their larger brains, humans continued to advance by making blade-type tools, cave paintings, developing sedentary communities, and eventually domesticating plants and animals.

Now let's apply this to Sasquatch. Eyewitnesses describe Sasquatch as a large creature. Given the massive size of a Sasquatch, it seems safe to say that they too have an enlarged brain—at least bigger than other primates. This statement is also supported by claims of other Sasquatch researchers who state Sasquatch are extremely intelligent. This is why they are so hard to find! So, in five million years, while living side-by-side with other hominids, Sasquatch has been unable to figure out how to make the simplest of tools or how to show any other example of being an advanced creature. That seems rather unlikely!

If Sasquatch is an early human species, then why don't we find evidence of Sasquatch villages? Humans are a species of community. Even early Neanderthals had lived in small, family-centered communes. I have heard the claim that Sasquatch only live in small groups, and that likely makes it easier for them to stay 'hidden' from humans. That would not be entirely different from Neanderthals, and yet we still don't find evidence of Sasquatch. But we do find evidence of Neanderthal dwellings that are hundreds of thousands of years old.

Hundreds of Native American villages have been discovered across the United States. Archeologists and Native American historians have easily identified which tribes established the villages, when the villages were active, and much more. So why hasn't anyone discovered an old Sasquatch village? Pre-historic or recent? It would seem logical that at least one or two villages, somewhere, would be a Sasquatch village—if Sasquatch are in fact a species of human. How odd.

If you are a fan of the cartoon *The Far Side* you are aware that Gary Larson often incorporated cavemen into the cartoon. One of my favorites depicts two cavemen attempting to fly a kite. One caveman is holding a log with a string tied to it. The other caveman is at the end of the string and trying to run. The first caveman is telling the other to run faster, because the 'kite' isn't lifting into the air. I bring up this cartoon to make a point. We see the cartoon and chuckle because we are familiar with the extremely primitive tools, or lack of tools, that cavemen had hundreds of thousands of years ago. Even though known 'cavemen' lived that long ago, it does make it rather difficult to accept that throughout those years, the greatest advance of a Sasquatch has only been improving their ability to hide in the woods.

I have heard some Sasquatch researchers claim that Sasquatch has evolved to live in nature, to commune with nature, and to be a highly intelligent being. Ok. But if that is true, then why do so many encounters with Sasquatch occur at night as they pass in front of a vehicle on a lonely road? Again, if the Sasquatch truly wanted to be undetected, they should wait a couple of seconds for the vehicle to drive by, and then cross the road. If they did, then no one would know that a Sasquatch was nearby. And the number of Sasquatch reports would drastically shrink. It's almost as if the Sasquatch wants to be seen. How odd for an *elusive* creature!

Chapter 5 | An-Alien-From-Outer-Space Theory

Another suggested theory for the explanation of Sasquatch is that Sasquatch is an alien species from another planet. To a large degree, this theory has developed from a number of encounters that involved other unexplained circumstances. For a growing set of eyewitness encounters with Sasquatch, the witness observed some number of flying balls, or orbs, in the close vicinity of the Sasquatch. These encounters vary in the number of orbs seen, the size of the orbs, and the amount of interaction between them and the Sasquatch. Some reports mention a single orb, while others may involve up to a dozen. The orbs vary in size from approximately baseball size, up to beach ball size. And the amount of interaction with Sasquatch is reported to be anything from the orb simply following the Sasquatch through the woods, to the orbs circling and spinning around the Sasquatch like a swarm of thirsty mosquitos, but the Sasquatch doesn't swat them like I do with mosquitos.

So, what are these orbs? Why are they observed with some Sasquatches and not others? They obviously are not the Sasquatch spaceship, as no reported Sasquatch could possibly fit inside a beach-ball-sized orb, and especially not the smaller orbs. And since these orbs are not spaceships, then what are they?

Since we're on the topic of spaceships, I have not yet found a reported encounter or sighting with a Sasquatch where a spaceship was observed—not even anything that could be misconstrued as a spaceship. So, if a Sasquatch is an alien, how do they get to Earth? The only reports that I have heard involving Sasquatch and aliens came through a friend of a friend of a friend, and so on. So I can't give that much credibility.

One common observation from every Sasquatch encounter that I have read or heard about, is that Sasquatch has the body size and shape description that I previously mentioned. That is of a large, hairy, humanoid-looking creature. I have not read or heard of any encounter or sighting in which the Sasquatch was wearing anything—no technological watches, helmets, intercoms, etc. . . . Not even clothes. I also have not heard of an encounter or sighting of Sasquatch where they were carrying anything other than a rock, stick, or another Sasquatch—nothing that could be considered 'high technology' that might aid them in space travel. So again, if a Sasquatch is an alien from another planet, how do they get to Earth without the assistance of some kind of technological aid?

And speaking of technology, let's address that issue. As I have pointed out, Sasquatch descriptions are of large, hairy human-oid-looking creatures. Sasquatch is described as being muscular, heavy-bodied. I have often heard, "It looked like a man, but much bigger." How does a creature with thick, massive fingers and appendages develop the highly sophisticated technology needed for space travel without the ability to make the simplest tools? In the evolution of humans, until we started living in a civilized community (i.e. people living and working together for the common good) everyone needed to be a big, strong hunter to survive. In community, smaller individuals with better dexterity were able to make clothes, tools, and even art. Those smaller individuals were able to advance civilization, and maintain the genetics of smaller individuals within the population. And we have already established that probably Sasquatch does not live in a community.

If you are a person of larger stature, you have likely experienced the frustration of trying to work with small pieces and parts. Perhaps you had to replace a tiny screw on the arm of a pair of glasses, or something like that. You may have had so much difficulty that you resorted to asking your spouse, child, or another family member for help. Or maybe you are of smaller stature, and you were asked to help someone because you have smaller hands and could more easily manipulate those tiny pieces back into place.

Another issue I want to address on the lack of technology is this: Sasquatch doesn't have opposable thumbs. I have watched many documentaries, read reports and listened to people recount their stories. One important fact is that even though the Sasquatch has five fingers, they do not have an opposable thumb. All five fingers on a Sasquatch hand are parallel. This would make it difficult to pick up and manage small objects, let alone build advanced technology and circuitry for a spaceship to travel across the universe.

So I have to ask: How does Sasquatch have the technology that would allow them to travel through space, from one planet to another, and to do so frequently? Apparently many of them enjoy regularly traveling to Earth on a regular occasion. Think of the rewards miles they must be racking up!

We're familiar with the 'little green men' description that has become the signatory life form for an alien in our present-day culture. Of all the documentaries, television shows, podcasts, and websites that I have viewed over the decades that contain eyewitness reports of people claiming to have seen an alien, I can't help but point out one peculiar observation: That little-green-man description often comes quite close to the physical descriptions given by eyewitnesses for what they observed.

It can be argued that the image of little green men as aliens has influenced people's perception of what an alien may look like. Be that as it may, the consistency is uncanny. There may be minor differences here and there. In some instances, the alien may be a little taller than E.T., but more often than not, that is how aliens are described. I point this out because the physical descriptions of 'aliens' are a stark contrast to the physical descriptions of Sasquatch, so much so, that if you with all seriousness approached anyone who claims to have seen an alien, showed them an artistic rendition of the typical Sasquatch, and then asked, "Is this what you saw during your alien encounter?" I'm quite certain they would laugh in your face. Or slap you in the face for not taking them seriously. I say this only half-jokingly because it is difficult to accept this theory that Sasquatch is an alien from another planet, when nearly all other evidence for aliens is in such contrast to what has been reported about Sasquatch.

I would also like to mention something about aliens from other planets. The Book of Genesis is the first book of the Bible. It is also the first book of the Jewish Torah. The Book of Genesis starts out telling the reader how God created the Earth, the Universe, and humans. Bear in mind that the Bible is the most reproduced piece of literature in the world. Regardless of which translation you read, the Book of Genesis clearly states that God created all life on Earth, and God created the moon and the stars to provide light during the darkness of night. It does not say that God created life in the 'universe.' Nor does it state that the multitude of stars and planets were created for the purpose of other life. So, just to be clear, Judeo-Christian Orthodoxy does not support the existence of aliens, and therefore the theory that Sasquatch is an alien from another planet goes against Judeo-Christian Orthodoxy.

Chapter 6 | An-Interdimensional-Being Theory

Another theory for the existence of Sasquatch that has grown in popularity in recent years goes even further than the alien theory. This theory claims that Sasquatch is an interdimensional being. For this chapter, I am addressing the version of this theory based on the concept that Sasquatch is something from a science fiction novel, and that Sasquatch can move from another parallel universe to ours, and back again.

This theory has taken root from a handful of Sasquatch encounters where the eyewitness saw the Sasquatch disappear. According to the eyewitnesses, they saw a Sasquatch at some distance, not too far from them; then in the blink of an eye, the Sasquatch quite literally disappeared. I have even heard of another Sasquatch encounter where the eyewitness claimed they saw the Sasquatch walk through a portal. Their description sounds like something from a science fiction movie.

Another part of the popularity of this theory is that it can provide some explanation to several of the conundrums that I have brought up throughout this book,—most notably the lack of physical evidence. If Sasquatch does not live on Earth full time, but is only a visitor, then it would be easy to conclude that Sasquatch individuals would rarely, if ever, die while visiting our dimension. And thus no Sasquatch skeletons or carcasses are to be found in the woods or under construction sites, or anywhere on Earth for that matter.

This theory would also help explain why we have not been able to catch a live specimen to place in one of our zoos for everyone to see. When a Sasquatch is being pursued by a would-be hunter or trapper, the Sasquatch would merely transport itself back to its home dimension to elude the pursuer. Even if a Sasquatch got caught in a trap, it wouldn't stay there for long as it would transport itself away to its home dimension.

This theory would also explain the randomness of their footprint trails. As I have mentioned, the footprint trails that people have collected cast molds from, seemingly suddenly start and suddenly stop. If a Sasquatch is from another dimension, and they arrive in our dimension through some type of inter-dimensional door, then it could be argued that they arrive through one door, walk some distance, and then return to their dimension through another door somewhere else. Thus the footprint trail would start at the point the Sasquatch entered our dimension, and the trail would stop at the point the Sasquatch left our dimension.

This theory would also explain why we haven't found any Sasquatch villages. If Sasquatch is only present in our dimension for a short period of time, they would not need to develop any kind of permanent living arrangements for themselves while they are here.

When you go through the Sasquatch encounter reports, especially the ones that occurred during the daytime, one commonality is that the Sasquatch was standing in the brush and watching the observer. As the Sasquatch was standing there, the observer would also describe how it was swaying back and forth. Sometimes the observer will state that the swaying motion is why they noticed the Sasquatch. So why do Sasquatch frequently sway back and forth? A theory I heard recently is that this habit is directly related to a Sasquatch being 'in tune' with the physical world. This is based on the notion that people with autism, or at least are on the spectrum, have trouble standing still. They sway back and forth. Their argument is that people with autism are more observant when it comes to others' actions, emotions, and motives.

Therefore, since autistic people are more 'in tune' with the people around them, and they often sway back and forth, then Sasquatch is more 'in tune' with nature and their interdimensional abilities, and that is why Sasquatch sways back and forth. Now, I have met quite a few people with autism. I can see how gifted they are. However, I have not met an autistic person that has the ability to travel between dimensions—at least not anyone that isn't a fictional character from a comic book.

In my experience, the most common argument against this theory is that a paranormal Sasquatch isn't possible, scientifically speaking. So I want to first address this misconception. Paranormal is a term to describe that which is outside or beyond what is currently known. Too often, people use this term when they think something is not physically possible, at least not as we know it to be. It is not impossible for a Sasquatch to be a paranormal entity that can travel from one dimension to another, but it is improbable in this context.

One argument against this theory for the existence of Sasquatch was mentioned in the previous chapter on aliens. The physical descriptions of Sasquatch do not infer their ability to create the needed technology for something as complex as space travel, nor the even greater complexity of inter-dimensional travel.

Also mentioned in the previous chapter, Sasquatch has not been observed to possess technology.

If you went to a nearby university, you would hear physics professors teaching that through relatively recent advancements in quantum physics, physicists have hypothesized the potential existence of other dimensions. Some hypotheses suggest anywhere from ten or more different dimensions are capable of existing along with our own. The amount of large equipment that is used in the experiments that brought the scientific community to these theories is quite impressive.

Now I have to ask this question: Given the size and complexity of the equipment necessary to detect the presence of another dimension, how much more sophisticated would be the technology required to achieve the *separation* of these dimensions? Outside of a Hollywood movie where you can imagine anything, the evidence of technology needed to regularly move objects and living creatures from one dimension to another and back again, is overwhelmingly complex. So now, how am I supposed to believe that a Sasquatch, a creature known for throwing rocks and sticks in an attempt to scare people away, has the necessary technology for inter-dimensional travel?

In some of these multiple dimension theories, it is suggested that the laws of physics across our universe in our dimension could possibly be different from the laws of physics in other dimensions. Living organisms may have abilities that we don't see in our dimension. Could this then explain the lack of technology accompanying any single Sasquatch? Because they don't need technology, they merely have the physical ability to move from one dimension to another? However, if that is the case for Sasquatch, then riddle me with this. If a Sasquatch is in one dimension with a certain set of laws of physics and a certain set of abilities, one of which would allow it to move out of their home dimension, wouldn't the laws of physics in the new dimension be the governing laws of physics? Thus a Sasquatch would not be able to transport back to its original dimension. Then it would find itself stuck in our dimension. But that doesn't seem to be what's going on with Sasquatch given the dilemmas addressed in the previous chapters.

Chapter 7 | A-Supernatural / Spiritual-Entity Theory

The last theory that I want to discuss about the existence of Sasquatch is this: A Sasquatch is a supernatural or spiritual entity. This theory has not garnered a lot of support, and yet it is the most common theory among Native American tribes. In many Native American cultures, Sasquatch is not a 'run-of-the-mill' supernatural being, like a magical fairy tale monster. Instead it is viewed more as a deity, something similar to those in Greek and Roman mythology. Given the descriptions of some of those mythological creatures, I can see where this theory could be attractive. However, with the lack of physical evidence of Sasquatch, there isn't much other supporting theoretical evidence for this theory outside of legends and folklore.

However, a couple of viewpoints taken from a Judeo-Christian perspective can account for the same phenomena, and provide some theological evidence for Sasquatch.

The first perspective from Judeo-Christian Orthodoxy surmises this: After God created Adam and Eve, as humans began to reproduce throughout the Earth, angels were assigned to watch over them. They were called the Watchers. Eventually, some of those Watchers came down to Earth and took humans as their spouses and produced offspring. There are also writings that claim the Watchers defiled wild animals. These offspring from the Watchers are referred to as the Nephilim, and described as giants. They are first mentioned in the Bible in the Book of Genesis. That book states that they existed before and after the flood for which Noah built the Ark.

They are mentioned throughout the Bible in the Books of Deuteronomy, Joshua, Numbers, Second Samual, and First Chronicles. There were the Anakites, descendants of Anak, who came from the Nephilim and lived across the Jordan River. The Rephaites, descendants of Rapha, who lived in Bashan are called 'giants.' There were the Amalekites that lived in Negev. Then there were the Hittites, Jebusites, Amorites, Emites, and Horites. And we can't forget the Gittites from Gath that often fought in wars alongside the Philistines. They are also written about more extensively in other ancient texts. The nine-foot-tall giant (sound familiar) known as Goliath was from Gath, before meeting his death at the hands of David. Could Sasquatch be the descendant of offspring of the Nephilim? Let's look at this a little deeper.

If Sasquatch are descendants of the Nephilim, much like Goliath, then would they be supernatural or mortal? If Sasquatch are the descendants of the Nephilim, and they are mortal, then the dilemmas discussed in Chapters 3 and 4 would still be issues.

Where are their physical remains? We know from the Bible that Goliath was mortal. At least enough so that he was killed with physical weapons. However, we must bear in mind that David was the anointed would-be king of God's chosen people. If Goliath was supernatural, it can easily be argued that divine intervention caused David's ability to kill a supernatural being. If Goliath was supernatural and had the same origins as current-day Sasquatch who are from the Nephilim, then it could be argued that Sasquatch would be supernatural too.

A rather demonic analysis is the second perspective of Judeo-Christian Orthodoxy to consider. If you have the opportunity to talk to an exorcist priest or a deliverance minister, or if you read a book on special deliverance prayers, you will see them refer to various demons, curses, spiritual enemies, demonic entities, and evil spirits. Of all the evil spirits, you may hear a prayer specifically against spirits of the earth, air, water, fire, the netherworld, and the satanic forces of nature. Could these spirits of the earth include Sasquatch?

Where do encounters with Sasquatch typically occur? In Nature! And isn't the netherworld in essence another dimension? Remember in the previous chapter when I discussed the hypothesis of multiple dimensions? From the Judeo-Christian Orthodoxy, a Sasquatch could possibly be an inter-dimensional, supernatural spirit or demon. Perhaps Sasquatch is a physical manifestation of a supernatural spirit or demon that is temporarily in the Earthly realm. After all, what is quantum physics? It's the idea that invisible forces influence the physical world. And it so happens that

Spirituality is this: invisible forces that influence the physical world. Some coincidence!

At this point I will address the possibility of Sasquatch as an interdimensional being from a supernatural perspective. Could there be evidence to support this idea? Let's look at the evidence.

Remember when I discussed the lack of photographs and video? The camera/video equipment that has been used for outdoor recreation, remote monitoring of property, and outright Sasquatch hunting is, and has for decades been, almost entirely electronic. Roger Patterson and Bob Gimlin carried a big, heavy film camera through the woods to get their famous thirty-second video more than fifty years ago. For the most part, few people bother to drag big, bulky, and heavy film cameras out into the wilderness. To be frank, people are lazy, and carrying heavy cameras into remote areas is hard work. Since cameras/video recorders went digital, the technology has become increasingly lightweight and extremely mobile. Today, trail cameras use light beam triggers; wireless cameras use Bluetooth connections to send video anywhere a signal can go; and personal body cameras record on small memory cards. Here is a question to ponder. Has anyone ever set up a Polaroid camera with a string for a tripwire, that is connected to a pulley, that is connected to a tube with a marble, that is connected to something else, and so on and so on, like an ultimate mouse trap for a low-tech attempt to trick a Sasquatch and get a clear picture? I think more of this kind of ingenuity has been employed by kids trying to get a picture of Santa Claus than has been used to get a picture of Sasquatch. I ask this only half-jokingly as Patterson and Gimlin have proven that older film cameras can capture a cleaner, clearer image of a Sasquatch, and to a much lesser extent, electronic cameras cannot.

Have you ever watched or read anything about ghost hunting? If so, you would be familiar with a common phenomenon that occurs in the presence of ghostly apparitions—electronic equipment goes haywire. It is undetermined if 'ghosts' disrupt electronics purposely, or if their non-physical physical form is in some way non-conducive to the flow of electrons that make electronics work. Perhaps their presence is merely enough to prevent electronics from working properly; no one knows. Regardless, this phenomena occurs; cameras go berserk, and no one gets a good photograph or video of a ghostly presence.

In Chapter 3, I discussed the cameras and video surveillance that could have, and should have, captured images or video of Sasquatch, but it hasn't worked. As I also mentioned, that has been electronic equipment. If Sasquatch is supernatural, as a fallen angel or a demonic spirit, in some similar manner that 'ghosts' are supernatural, then could this same phenomena about electronic disruption explain why we haven't seen quality photographs or videos of Sasquatch in over fifty years? Could the presence of a Sasquatch prevent people from collecting quality video or photographs of Sasquatch?

So what about the lack of physical evidence?

In Judeo-Christian Orthodoxy, Heaven is considered to exist as another dimension from this earthly realm where immortals currently reside. As I mentioned in the previous chapter, physicists now agree that the existence of other dimensions is scientifically possible. If God, His angels, and the past souls worthy of entrance are in a heavenly dimension, shouldn't Satan, his demons, and the unworthy souls be residing in a dimension of hell?

There is no active collection database for reports of angelic interactions that I am aware of, at least not on a level equal to that of the BFRO website for Sasquatch encounters. Reports of angelic interactions include physical interactions, but also include hearing voices, visits in dreams, and helping people remember past memories. It can be argued whether an interaction is with an angel or a demon, so the person that received the 'visit' must discern which their visitor was. An angelic visit and 'message' will contain biblical truth; it will have a distinguishable benefit, and there is no violence against the individual. Anything other than that would be demonic. A number of documentaries, television shows, and other media outlets have described angels physically intervening in people's lives. Numerous examples and stories have been collected from around the world involving a miracle, and the 'hero' of the story is accredited to being an angel. I'm sure you've heard some. They tend to go something like this. A person was going about their normal, daily activities; then suddenly, tragedy strikes; another person shows up seemingly out of nowhere; the stranger helps the person in dire need; then this 'hero' seemingly disappears as quickly as they arrived. When these recipients of angelic intervention talk about their situation, the angel appeared as a real person. The angel wasn't ghostly, as in movies. In fact, if the circumstances of the situation were not so perilous, the encounter would have been completely ordinary. So, if supernatural beings like angels are able to move from Heaven to Earth and back (in other words be inter-dimensional), then isn't it plausible that a Sasquatch who is a fallen angel or demonic spirit can be inter-dimensional also?

If angels and demons are both supernatural and interdimensional, then perhaps the laws of physics regarding multiple dimensions are not a limiting factor of their abilities. It is now plausible that they could have the ability to move from one dimension to another and back without an issue. After all, that is in essence what 'super' natural means—being greater than what normally exists in the natural world.

If a Sasquatch is an inter-dimensional, fallen angel or demonic spirit, then that could account for the lack of physical evidence found on Earth. The Sasquatch would spend much of their time in another dimension—like hell or somewhere else—shift into the Earthly dimension for whatever reason, then shift back to wherever they came from. This would explain the lack of deceased bodies, tracks starting and stopping out of nowhere, and a lack of villages or other living quarters.

As I have mentioned earlier, Sasquatch sightings occur across North America. How can something be populous in forty-nine states and nine provinces of North America, and yet leave no evidence besides an occasional set of footprints? And they do that without being a supernatural, interdimensional being?

Remember in Chapters 3 and 4 when I mentioned the large number of Sasquatch encounters that occur at night on backroads and lonely highways? I also mentioned that if the Sasquatch truly wanted to remain elusive, they would wait a few seconds for the vehicle to drive by, then cross the road. But they don't. They cross the road in front of the vehicle, allowing the driver and the occupants to 'see' the Sasquatch. What is even more perplexing about this isn't just that the Sasquatch crosses the road in front of the vehicle, but that the Sasquatch crosses the road at a moment when they can be viewed by the driver, and yet have just enough time to get out of the way of the vehicle so as not to get hit. As mentioned earlier, I am aware of no report describing a Sasquatch becoming roadkill. That is impeccable timing, time after time!

Here is the reason I mention this oddity now: As people report these sightings that occur at night in the headlights of their vehicle, it is the follow-up that I find intriguing. Many people go on to say that it changed their lives—and not for the better. They have trouble sleeping. They don't go out at night anymore. They avoid that road or area. For most people, a Sasquatch encounter has a very negative impact on their lives—even an encounter so brief as a Sasquatch crossing in front of their vehicle. Does that sound like the works of a primate or Neanderthal? Or does that sound like the works of a demonic entity?

I mentioned in Chapter 3 that Sasquatch's feet are adapted to walking upright as a large and heavy creature. I also mentioned that a lot of evidence in the cast footprints supports this. Then, I ask again, how are Sasquatch able to move as fast as they do, without being supernatural?

I also mentioned in Chapter 3 that Sasquatch are sometimes

seen climbing in trees. I bought this up for the simple fact that Sasquatch are extremely large, and extremely heavy creatures. And as I asked before, how does a creature that large climb around in trees without breaking limbs and falling from the tree? The tree branches have to be extremely strong!

Remember when I mentioned that a large number of Sasquatch sightings involve the observer becoming overwhelmed with fear? The Bible describes many occasions when an angel of God appears to an individual, and that person is overcome with fear—for instance, Mary before she conceived Jesus; Elizabeth and Zechariah before they conceived John the Baptist; the shepherds watching the flocks the night Jesus was born, and Simon and John when they accompanied Jesus during the Transfiguration—just to name a few. In each of these situations, as well as the other instances mentioned in the Bible, the following sequence occurred. The angel manifested in a physical form; the observers were overcome with fear, and the angel responded telling them not to be afraid. Then the fear subsided, and the angel delivered their message.

When people reported an encounter with a Sasquatch and were then overcome with fear, did the Sasquatch tell them not to be afraid? An occasional grunt or scream is the only Sasquatch response that I have heard of when an unsuspecting witness is overcome—or not overcome—with fear. This is hardly the kind of response one would expect from a 'good' supernatural being. But this might be something that one would expect from a fallen angel or a demonic spirit—that the Sasquatch/fallen angel/demonic spirit would 'want' you to remain fearful.

It should be noted that many other instances in the Bible describe an angel appearing to someone, and that person does not experience overwhelming fear. However, in those situations, the angel was there for something good. Can that be said of other Sasquatch encounters? What 'good deeds' has any Sasquatch done for anybody?

You may be familiar with these Bible passages. If you are not familiar with the Bible, you may not realize that an equal number of verses talk about demons and angels. It is surprising to me that many Christians are willing to accept the possibility of angels having an influence in someone's life, but they are reluctant to accept the possibility of demons having an influence in someone's life.

I mentioned earlier about people being overcome with fear, and I mentioned the 'fear' response of pets (most notably dogs). Why do dogs and other animals respond differently to Sasquatch? I have often heard people say that animals have a sixth sense. Sometimes people will say something like, "My pets can tell when the weather is about to change," or "My dog can tell when someone is sick, even if that person doesn't show symptoms." Could it be that an animal's sixth sense allows them to see the Spirit world? Or maybe at least they can see a Sasquatch for what it is?

Along with pets and dogs having their own fear response to the presence of Sasquatch, the other critters in the woods seem to have their own response. Remember when I mentioned that people reported the woods becoming eerily quiet? It seems that birds and other woodland animals might be able to 'see' Sasquatch as something more than just a primate.

In the previous chapter, I mention that Sasquatch tends to sway back and forth. In many encounters, especially those that occurred during the day, the observer noted that the Sasquatch was standing and swaying from side to side. I mention this again for a couple of reasons. First, for a creature that supposedly wants to be hidden, it is not wise to stand *beside* a tree (instead of *behind* the tree) and move from side to side. In fact, many observers state that the swaying motion is why they saw the Sasquatch in the first place. Had the Sasquatch not been swaying from side to side, they likely would have gone unseen.

Here is the second reason for again mentioning this habit: Some argue that Sasquatch's swaying from side to side is similar to autistic people. But other people have this same habit. If you have ever been in the presence of a deliverance minister as they are casting demons out of people, or even just watched videos of this occurrence on the internet, you would notice something peculiar. To clarify, most people are not demonically possessed—where demons have taken complete control of a person's body. In contrast, most people seeking deliverance are demonically *oppressed*. This is when the demons have a heavy influence on the person—their thoughts, words, and actions. So, when a person is receiving deliverance, and the demons within manifest themselves, that person frequently begins to sway back and forth. This is usually before they have convulsions and then vomit. But my point is that when a demon begins to reveal itself, the person often begins swaying from side to side. I mentioned earlier that Sasquatch has a habit of swaying side to side when it reveals itself to people. Could it be that Sasquatch is some kind of demonic manifestation?

At the end of Chapter 3, I discussed the inconsistencies of Sasquatch behavior. Sasquatch sightings rarely, if ever, occur at the same location. I also mentioned that most people will rarely get a response from a wood knock, the reported go-to communication method of Sasquatch. I mention this again because I have not had the time or resources to collect data on every Sasquatch encounter concerning, date, time, weather conditions, cloud cover, moon phase, habitat conditions, etc. To do so might give one an idea of when Sasquatch encounters occur, but that is a big if. I am more inclined to wonder if the Sasquatch encounter is less about the where and when, and it is rather about *the who*. If Sasquatch is some kind of demonic entity, then it is much more likely that the encounter has more to do with the witness than with the Sasquatch or the location. Until I, or someone else, have time to do an extensive survey of everyone who has had an encounter with Sasquatch, this issue will remain a theory. But given the fact that Sasquatch encounters are unpredictable, a survey of the witnesses is the next logical step.

As one searches the growing number of Sasquatch sightings and reports, they can find several reports from people that claim to have 'communicated' with a Sasquatch. No, I'm not saying that they sat down with Sasquatch at a picnic and had a deep conversation about the 'other' great mysteries in life (pun intended). As their reports explain, the Sasquatch was able to talk to the person through telepathy. This 'mind-speak' as they refer to it, is something like a warning system. The Sasquatch 'tells' the person they are nearby, behind a tree, or something similar, and then the person is not alarmed or frightened upon seeing the Sasquatch. I'm not an expert on the communication abilities of Sasquatch, but it is completely reasonable that they would have some type of unique communication style among themselves. After all, every species of animal can communicate with others of their species in some way. Whether it be vocally with various sounds, or from body language with various movements of their body parts, they can communicate with each other. And humans have created over 7,100 different languages (English, Spanish, French, etc.).

I mentioned the Sierra Sounds earlier. From that research, it is claimed that Sasquatch does have a language for speaking to other Sasquatch. You might be thinking that this could be evidence that Sasquatch is an advanced species. I have heard about several Sasquatch encounters where the observer reported witnessing multiple Sasquatches chattering back and forth to each other as if they were having a conversation in their own language. So, this is why I think this 'mind-speak' or telepathy is evidence that Sasquatch is supernatural:

How did these Sasquatch learn English?

If Sasquatch were a primate, or Neanderthal, or another human species that developed telepathy ability, and had a language to communicate with other Sasquatch, first, why are they able to use that ability on humans, and secondly, why are we able to *understand* what they are saying? Shouldn't that message come across as the same gibberish chatter that is heard on the Sierra Sounds, or is similarly described as "chipmunk chatter" that other observers have witnessed? Not only are the messages understandable; they seem to have grammatical structure. Apparently, Sasquatch doesn't speak like Tarzan (Me Sasquatch, you Jane!), but can speak in complete sentences. Another question that makes this whole thing perplexing: Have any of these recipients of 'mind-speak' messages bothered to respond to the Sasquatch and ask, "What are you?" It seems like that would be the perfect opportunity to get a definitive answer!

Chapter 8 | A Warning

After reviewing each of these theories in depth—based on the lack of evidence, or inconsistencies in the available evidence—it's hard to accept the notion that a Sasquatch is just another great ape, a Neanderthal, alien, or science fiction-style interdimensional traveler. At least we need additional evidence to answer these looming questions. At this time, the most likely theory says that Sasquatch is a spiritual being.

Whether a Sasquatch is some kind of fallen angel offspring or physical demonic apparition, or something else altogether, it would be prudent to proceed with caution when continuing to investigate or encounter Sasquatch.

In the previous chapter, I discussed angels appearing to human beings. I mentioned that in every report that I have found, heard, or read about involving appearances of angels, the witness either didn't have an overwhelming sense of fear, or they were comforted by the angel to remove that fear. I also mentioned how this is a stark contrast to the encounters with Sasquatch. If you're not familiar with Judeo-Christian Orthodoxy, angels are inhabitants of Heaven, and they serve God's will. They are inherently good, and do good things at God's command. This is why we hear about them delivering good news, or doing good works to help people in need.

I have not heard of a Sasquatch encounter where the Sasquatch did something good for the witness. I am familiar with many encounters where the Sasquatch was angry, threw large rocks and logs at people, and even attacked the witnesses. This certainly isn't the behavior of an angel. I bring this up again for the following reason. If Sasquatch is a supernatural being, there is no supporting evidence that they are good supernatural beings. In several Native American cultures, people would give offerings of food to the local Sasquatch. If the offerings were not regularly provided, the Sasquatch would begin to 'terrorize' the tribe.

Certainly, not every reported Sasquatch encounter is a terrifying event, where the witness narrowly escapes with their life. And not every witness is negatively impacted by their encounter. But that should make you wonder what the Sasquatch is trying to do. What is it up to? What is its motive for 'letting' you witness their presence? If angels appear to people to do good, then wouldn't it be likely to presume that a fallen angel or demon would appear to people to 'not' do good?

I mentioned in the previous chapter that many people who have witnessed Sasquatch tell of the anguish that proceeded in their lives. People have stated that they have trouble sleeping at night. Or they don't go outside at night. Or they avoid certain areas or places altogether. But the bottom line is that these people have their lives changed for the worse, not for the better. And many of these people caught only a passing glimpse of a Sasquatch from inside a vehicle. Again, that doesn't sound like the work of a primate or Neanderthal. But it does sound like the work of a demonic entity.

If you talk to an exorcist priest, a deliverance minister, or if you see a video of demons being cast out of a person, you will learn that the demonic entity that is possessing that person has evil intentions. It might be trying to get the possessed person to harm themself or others. It might even be trying to 'ruin' that person's life. Whatever the intention of the demon, it's not good for the possessed.

If you are out hunting, on a hike, or driving down a road, and you witness a Sasquatch, be careful. If the Sasquatch appears angry; or is throwing rocks at you; then it is obviously trying to harm you, or is *about* to harm you. If the Sasquatch doesn't appear angry, and at the moment seems calm, or even 'interested' in who you are and what you are doing, then you have to wonder what the Sasquatch is up to. Ultimately it is not in your best interest.

I have come across several stories of people interacting with 'ghosts.' I'm sure you have heard about some of these, as they have been incorporated into television shows and movies. The interactions I'm referring to involve someone having an encounter, and at first the ghost doesn't do anything destructive or harmful, so they become 'friends.' Then later, the friendship sours, and the ghost becomes increasingly harmful. The encounters I am familiar with typically involve children, as their youthful naivety and desire for friends makes them unaware of the spiritual danger they are getting into.

You may have had an encounter with a Sasquatch that seemed 'innocent,' and therefore you are not bothered to venture out to the same location without worries of additional encounters. You may be someone who purposely goes to areas where other people have reported encounters with Sasquatch in hopes of having your own encounter. You may even be attempting to have multiple encounters so you can gather new evidence for your own explanation of Sasquatch. Whatever your reason for purposely putting yourself face to face with a Sasquatch, you must be aware that at some point, the 'innocence' will disappear. At that point, are you ready to find out what kind of evil the Sasquatch has intended for you?

In the previous chapter, I mentioned that the inconsistencies of Sasquatch encounters may have more to do with the witness than the Sasquatch. If this is true, then the witness wasn't just in the right place at the right time. The witness was supernaturally singled out to have that encounter. If the witness was then in fact singled out by the Sasquatch for their encounter, then you have to ask, "Why."

Here is something else I would like you to think about. If you are familiar with Judaism or Christianity, then you are probably familiar with the Ten Commandments. The first commandment says, "I am the Lord your God, you shall not have any other gods before me." This means that everyone must love God with all their heart, all their might, and all their soul. This is difficult. I say this because when people decide to go golfing instead of going to church, or they skip church to go fishing, then they are putting golf and fishing above God, and thus breaking the first commandment. So if you are spending your free time 'researching Sasquatch' and not making time for worshiping God, you too are breaking the first commandment.

I asked earlier, what evil does a Sasquatch have for you? It could simply be to subtly draw you away from God into condemnation. If you possibly fit into this situation, don't think I'm saying you cannot continue with your research. What I do want to do is make you aware of this ploy so that you can correct your situation. Once you correct your situation, and you put God back at the center of your life, be aware that the demons in your life, and maybe Sasquatch too, will change their plans for you as well!

References

Bigfoot Field Research Organization [BFRO]. 1995. BFRO home page < http://bfro.net>. Accessed on 13 April 2023.

Bindernagel, J.A. 1996. *North America's Great Ape: the Sasquatch.* Beachcomber Books, Courtenay, Canada.

Bookhout, T.A., editor. 1996. *Research and Management Techniques for Wildlife and Habitats.* The Wildlife Society, Bethesda, USA.

Deutscher, G. 2010. *Through the Language Glass.* Metropolitan Books, New York, USA.

Ember, C.R., M. Ember. 1993. *Anthology, Seventh Edition.* Prentice Hall, Englewood Cliffs, USA.

Ethnologue. 2023. Ethnologue home page <https://www.ethnologue.com/> Accessed on July 22, 2023.

Gorillas World. 2023. Gorillas World home page <https://www.gorillas-world.com/gorillas-in-captivity/> Accessed on 12 August 2023.

Green, J. 1978. Sasquatch - The Apes Among Us. Hancock House Publishers, Surrey, Canada.

Gribbin, J. 2020. The Many-Worlds Theory, Explained. <http://www.thereader.mitpress.edu/>. Accessed on 27 May 2023.

Guideposts. <https://www.guideposts.org>. Accessed on 13 May 2023.

Kisling, V. N. 2001. *Zoo and Aquarium History.* CRC Press, Boca Roton, USA.

Krantz, G. S. 1999. *Bigfoot Sasquatch Evidence.* Hancock House Publishers, Surrey, Canada.

Larson, G. 1984. *The Far Side Gallery.* Andrews and McNeel, Kansas City, USA.

Lighthouse Online. 2023. <https://lighthouseonline.com/blog-en/how-many-languages-are-there-in-the-world/> Accessed on July 22,2023

Meldrum, J. 2006. *Sasquatch: Legend Meets Science.* Forge Books, New York, USA.

New American Bible. 1968. Catholic Book Publishing Company, New York, USA.

Ripperger, C. 2018. *Deliverance Prayers For Use by the Laity.* Sensus Traditionis Press, Keenesburg, USA.

Schmuller, J. *Statistical Analysis with Excel for Dummies*, 3rd Edition. John Wiley and Sons. Hoboken, USA.

Smith, D.W, D.R. Stahler, and D.R. MacNulty. *Yellowstone Wolves.* University of Chicago Press. Chicago, USA. 2020.

United States Census Bureau. <https://data.census.gov> Accessed on July 29, 2023.

www.ingramcontent.com/pod-product-compliance
Lightning Source LLC
Chambersburg PA
CBHW072212270326
41930CB00011B/2618